CATERPILLARS Play Tricks!

BY JENNIFER LOMBARDO

ANIMAL TRICKSTERS

Gareth Stevens PUBLISHING

Please visit our website, www.garethstevens.com. For a free color catalog of all our high-quality books, call toll free 1-800-542-2595 or fax 1-877-542-2596.

Library of Congress Cataloging-in-Publication Data

Names: Lombardo, Jennifer, author.
Title: Caterpillars play tricks! / Jennifer Lombardo.
Description: Buffalo, NY : Gareth Stevens Publishing, [2025] | Series: Animal tricksters | Includes index.
Identifiers: LCCN 2023035538 | ISBN 9781538293492 (library binding) | ISBN 9781538293485 (paperback) | ISBN 9781538293508 (ebook)
Subjects: LCSH: Caterpillars–Juvenile literature. | Caterpillars–Behavior–Juvenile literature. | Animal defenses–Juvenile literature. | Camouflage (Biology)–Juvenile literature.
Classification: LCC QL544.2 .L687 2025 | DDC 595.7813/92–dc23/eng/20230831
LC record available at https://lccn.loc.gov/2023035538

Published in 2025 by
Gareth Stevens Publishing
2544 Clinton Street
Buffalo, NY 14224

Designer: Andrea Davison-Bartolotta
Editor: Jennifer Lombardo

Photo credits: Cover, p. 1 InsectWorld/Shutterstock.com; series art (background texture) Kriangkrai Namtongbai/Shutterstock.com; series art (caterpillar texture) Adrian Radulescu/Shutterstock.com; p. 5 BlueRingMedia/Shutterstock.com; p. 7 Piper333/Shutterstock.com; p. 9 Tyler Fox/Shutterstock.com; p. 11 Boox Selekta/iNaturalist.com; p. 13 Nancy Bauer/Shutterstock.com; p. 15 blickwinkel/Alamy Stock Photo; p. 17 Rudmer Zwerver/Shutterstock.com; p. 19 (main) IanRedding/Shutterstock.com; p. 19 (inset) Boris Riauzov/Shutterstock.com; pp. 20, 21 Jay Ondreicka/Shutterstock.com.

Printed in the United States of America

CPSIA compliance information: Batch #CSGS25: For further information contact Gareth Stevens at 1-800-542-2595.

CONTENTS

Words in the glossary appear in **bold** type the first time they are used in the text.

COOL CATERPILLARS

Caterpillars are amazing **insects**! They look like worms, but they're actually the **larval** form of butterflies and moths. When their egg hatches, or breaks open, they crawl out and start eating right away.

Caterpillars need to eat enough to keep them alive during the next part of their life. In this part, or stage, they close themselves inside a **chrysalis**. In the chrysalis, they turn into goo. Then the goo changes into the caterpillar's adult form—a beautiful butterfly or moth.

LIFE CYCLE OF A MONARCH BUTTERFLY

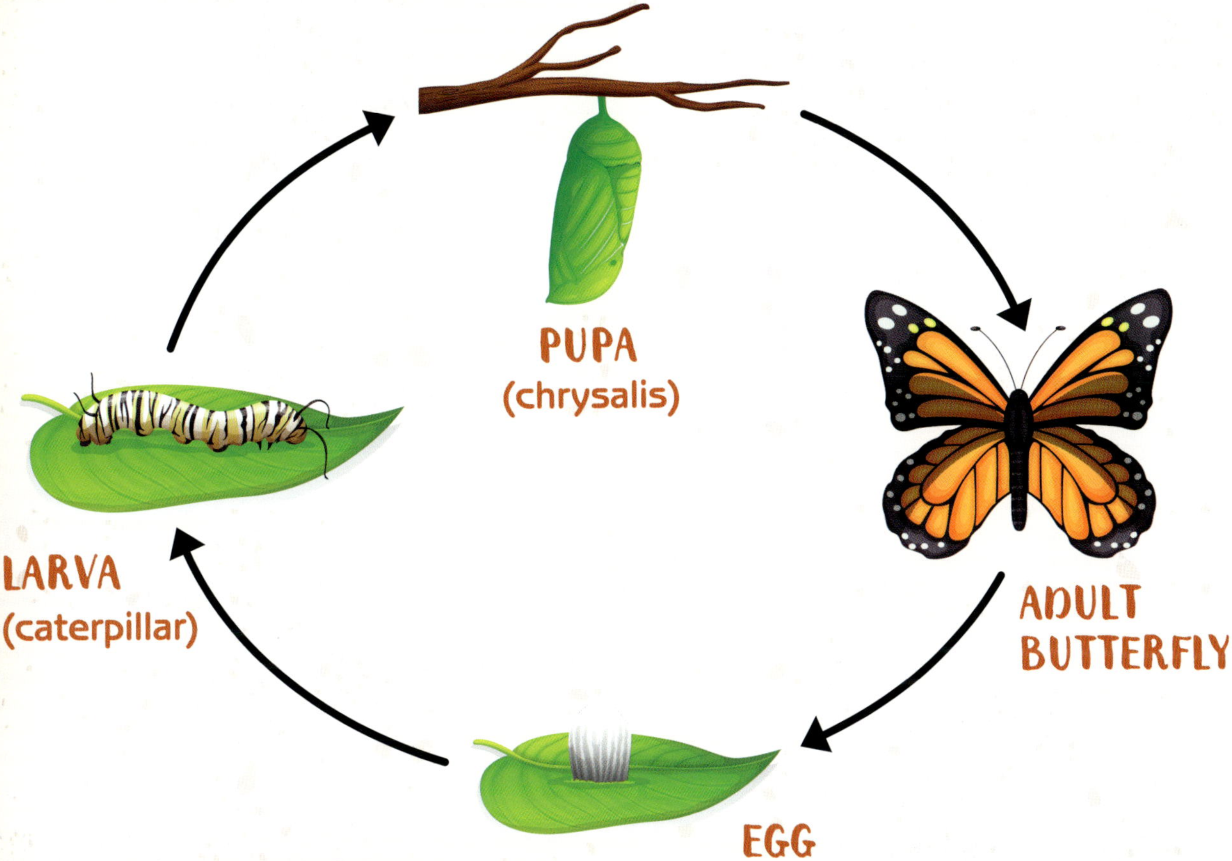

FUN FACT

Because caterpillars are the larvae of butterflies and moths, they're often called by their adult name.

This picture shows the life cycle of a butterfly.

THE MONARCH

Caterpillars don't have many **defenses**. They aren't good at fighting, and they don't move very fast. Because of this, they need ways to trick predators into leaving them alone.

Monarch caterpillars live on a plant called milkweed. Milkweed has a **toxic** liquid that makes most herbivores, or plant eaters, leave it alone. But monarchs eat it. They keep the toxin in their body. This makes them taste bad to some predators, such as birds. If a bird eats a monarch, it will throw it up! That teaches the bird not to eat monarchs.

This monarch caterpillar is eating a milkweed leaf.

FUN FACT

Monarchs are toxic only to vertebrates, or animals with a backbone. Spiders and ants can eat monarch caterpillars and eggs.

THE BLACK SWALLOWTAIL

The black swallowtail caterpillar's defense is **camouflage**. When it's very young, it looks like bird poop! As it gets older and bigger, it changes color. It makes itself look like a monarch caterpillar. This tricks the swallowtail's predators into thinking it tastes bad.

All caterpillars go through stages of **development** called "instars." By its fifth and last instar, a black swallowtail has lost all its camouflage. By then, the caterpillar doesn't need it anymore. It's ready to make a chrysalis.

This picture shows the black swallowtail camouflaged as bird poop.

THE GAUDY SPHINX

The gaudy sphinx caterpillar also uses camouflage. It has large fake eyes on its body. These are called "eyespots." Scientists think eyespots scare predators by making them think they're looking at a bigger animal.

The gaudy sphinx also has another trick. It can mimic, or copy, a snake! It's got the same colors on its body as some kinds of snakes. As it grows, its head also takes the same shape as a snake's head. Even people are sometimes fooled!

The gaudy sphinx can look just like a snake.

THE WOOLLY BEAR

The larva of the Isabella tiger moth has many names. It's often called the woolly bear, fuzzy bear, or woolly worm. Some people also call it the hedgehog caterpillar. This is because of its trick. When something picks up this caterpillar, it curls into a ball and plays dead. Scientists call this trick "thanatosis" (tha-nuh-TOH-sihs).

Most predators don't want to eat dead things. When the caterpillar plays dead, predators often leave it alone. Once they're gone, the caterpillar starts moving again.

People used to think the amount of black fur on a woolly bear would tell them how much it was going to snow in the winter.

FUN FACT

The woolly bear can stay alive in weather as cold as -90°F (-68°C)! It does this by freezing its body and unfreezing, or thawing, when it gets warm.

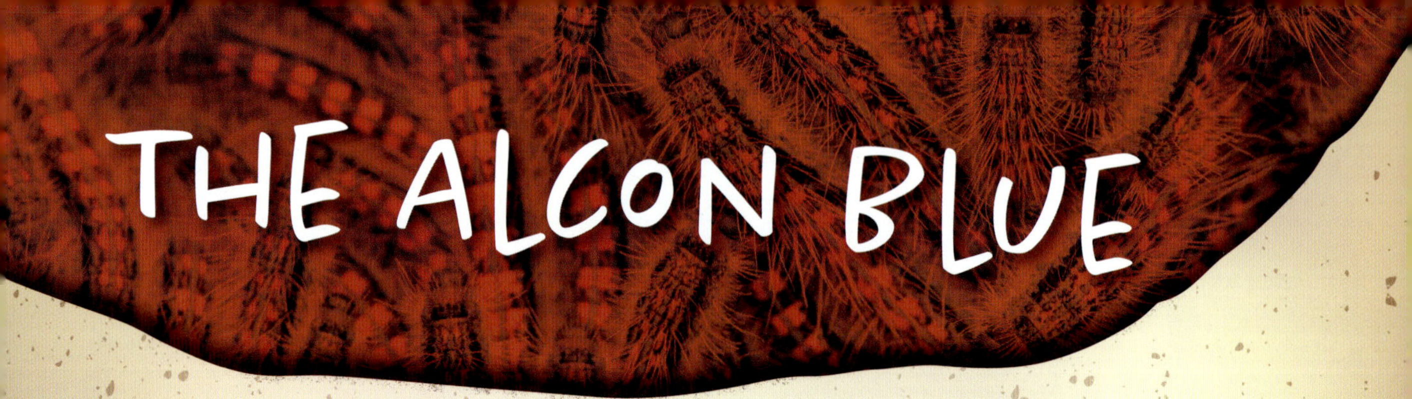

THE ALCON BLUE

Most caterpillars are herbivores. However, the larva of the Alcon blue butterfly is a carnivore, or meat eater. It feeds on red ants. This caterpillar has a tricky way of getting its food.

A few weeks after the caterpillar hatches, it falls to the ground. It waits there for nearby red ants to find it. The caterpillar gives off a smell that tricks the ants into thinking the caterpillar is an ant larva. The ants take it back to their home.

This picture shows an ant coming across an Alcon blue caterpillar.

LIVING WITH ANTS

For the next few months, the Alcon blue caterpillar lives like an ant. The adult ants bring it food. However, the caterpillar also eats the ant larvae! It starts with the biggest ones. It lets the smaller ones grow, then eats them too. This way, the caterpillar has enough food to last until it's ready to make its chrysalis.

When the Alcon blue butterfly comes out of its chrysalis, it leaves the ant nest. It flies away to lay eggs.

FUN FACT

The Alcon blue caterpillar can also make sounds that mimic worker and queen ants.

The Alcon blue butterfly feeds on plants, not ants.

HAWAIIAN CATERPILLARS

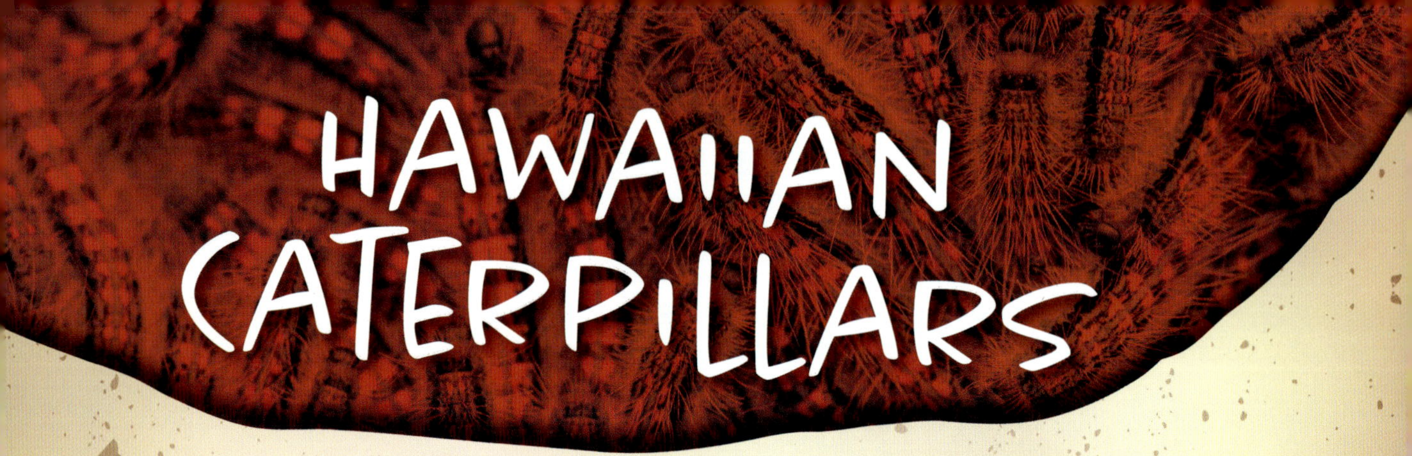

There are 20 species, or kinds, of moth larvae that are native to Hawaii. Of those 20 caterpillars, 18 are carnivorous! They use camouflage to hide from the insects they eat. Some are green because they hide on leaves. Others are brown to let them blend in with sticks.

The caterpillars have tiny, **sensitive** hairs all over their body. When an insect touches one, the caterpillar shoots forward to attack! It grabs its food before the insect can fly away.

When the caterpillar finds a bug to eat, it wraps its six sharp front legs around the insect.

THE SADDLEBACK

The saddleback caterpillar gets its name from the green spot on its back that looks like a **saddle**. It has spots on its rear end that look like a face. This tricks predators into grabbing the wrong end.

The caterpillar also has stinging **spines** all over it. These hurt many animals. They can also hurt people, even though we don't eat caterpillars! If a person touches a saddleback caterpillar, the spines will get stuck in their skin. People say it hurts worse than a bee sting.

This picture shows the "saddle" that gives the saddleback caterpillar its name.

FUN FACT

The saddleback caterpillar has sticky feet. This is why it's sometimes called a "slug caterpillar."

GLOSSARY

camouflage: Colors or shapes on animals that allow them to blend in with their surroundings.

chrysalis: A hard case that keeps a moth or butterfly safe while it turns into its adult form.

defense: A way of guarding against an enemy.

development: The act of growing and changing.

insect: A small animal with six legs and three main body parts.

larva: A bug in an early life stage that has a wormlike form. The plural form is larvae.

saddle: A padded seat used to ride a horse or mule.

sensitive: Able to sense or feel changes in surroundings.

spines: A stiff, pointed projecting part of a plant or animal.

toxic: Causing harm or death if eaten or touched.

FOR MORE INFORMATION

Books

Maurer, Tracy Nelson. *Caterpillars.* New York, NY: Crabtree Publishing Company, 2022.

Murray, Julie. *Spicebush Swallowtail Caterpillars.* Mendota Heights, MN: Dash!, 2023.

Schutten, Jan Paul. *Inside In: X-Rays of Nature's Hidden World.* Vancouver, CA: Greystone Kids, 2021.

Websites

BioKids: Butterflies and Moths
www.biokids.umich.edu/critters/Lepidoptera
Learn about the differences between butterflies and moths.

Kiddle: Caterpillar Facts for Kids
kids.kiddle.co/Caterpillar
Read more about different kinds of caterpillars and take a look at some cool photos.

National Geographic Kids: The Butterfly Life Cycle
www.natgeokids.com/uk/discover/animals/insects/butterfly-life-cycle
Learn more about each stage in the butterfly life cycle.

INDEX